Stones

CATHERINE CHAMBERS

RSVP
**RAINTREE
STECK-VAUGHN**
P U B L I S H E R S
The Steck-Vaughn Company

Austin, Texas

Published by Raintree Steck-Vaughn Publishers, an imprint of Steck-Vaughn Company

Library of Congress Cataloging-in-Publication Data
Chambers, Catherine.
    Stones / Catherine Chambers.
        p.   cm. — (Would you believe it!)
    Includes bibliographical references and index.
    ISBN 0-8172-4105-1
    1. Stone — Juvenile literature. [1. Stone.] I. Title. II. Series.
TA426.C47   1996
553.5 — dc20   95-33269
                                                              CIP
                                                               AC

Printed in Hong Kong
Bound in the United States
1 2 3 4 5 6 7 8 9 0 LB 99 98 97 96 95

Contents

What Is Stone?

Stone is all around us, from high, craggy peaks to small, shiny pebbles. Most of it was formed on Earth millions of years ago. Stone is often hard and rocky, but it can be soft and crumbly. Over a long period of time, it is worn down into sand or clay. Some stones contain precious gems. Others have useful minerals. All over the world, people use stone in unbelievable ways.

Crystal stones
This clear crystal rock is emerald, a gemstone. People will cut and polish it to make beautiful jewelry.

Soft rock
This layered rock on the coast of Wales was formed by layers of mud and silt being crushed together. It took millions of years to form. This sort of rock is called sedimentary rock. Its stone is the softest of the main rock types.

Hard rock

This hill in Zimbabwe is made of granite, a type of igneous rock. Millions of years ago, these igneous rocks formed when hot liquids bubbled up from the center of the Earth. They later cooled to become solid rock.

Changing rock

Changes on Earth or deep inside it affect rock. Some igneous and sedimentary rocks have been changed by extreme heat or pressure. The new rock is called metamorphic. This snow-covered slate is a type of metamorphic rock.

Shaping Stone

Stone can be carved, chipped, and smoothed into amazing patterns and shapes.

As soft as soap!

People have carved soapstone for thousands of years. It is not really as soft as soap, but it is soft enough to shape easily. This figure is from an ancient kingdom in Central Africa. It was made to stand on a grave.

Alabaster carvers

In Egypt, pearly vases are carved out of soft, light alabaster stone. The skilled carvers can make the surfaces very smooth. Many of these vases are sold to tourists.

Words in stone

On the Tibetan Plateau, a stonemason carves letters into stone tablets with a chisel. The words are prayers and teachings of the Buddhist religion.

Detailed carving

These delicate designs are carved on tombs in Pakistan. They are traditional patterns connected with the Islamic religion.

Stone statues

These stone bulls guard a palace in the French city, Marseilles. Statues and fountains are often carved from granite. This tough stone stands up well to harsh weather.

Grinding Stone

For thousands of years, hard stones have been used to grind cereals and spices. And rubbing clothes against a washing stone gets them really clean!

Pumice power

Pumice is solid volcano lava! The air bubbles inside it mean it is light, and it floats. Yet it can rub hard skin off your feet. Powder made from pumice is used for rubbing wood, leather, and stone smooth.

Fine flour

This sculpted millstone and bowl are used in Ethiopia to grind wheat grain into flour. Grain is crushed between the heavy stones as the millstone turns.

Stone sounds

Hitting rock can make a loud noise, but it also wears rock away. You can see holes worn into this old rock gong in Tanzania. These types of rocks were probably used as musical instruments or to give warning signals.

Clean clothes

Washing stones have been used in rivers all around the world for thousands of years. These women in Syria are using some. The stones are good for rubbing dirt off clothes.

Stone Buildings

Most of the stone on Earth is millions of years old — so we know it can last a very long time. This makes stone good to build with. From huts to castles, stone buildings are strong, safe, and built to last for a long time!

Safe houses
Rough stone chunks make the walls of this Italian house. The slate roof is topped with a chimney that is made of smooth stone blocks.

Strong shapes
The Church of the Sacred Heart in Paris was built over 100 years ago. The stones have been made into many shapes. The strong bell tower holds one of the largest bells in the world.

Ancient stones

This wall of huge stone blocks was built by the Incas of Peru. The blocks fit together perfectly—it is impossible to even fit a knife's blade in between them.

Cave castles

You walk through a narrow valley to get to El Kazneh, the Treasury of the Pharaoh. This gigantic building is part of the ancient city of Petra, in Jordan. Petra is known as the "rose red city," because its buildings are made from red sandstone rock.

Hot Stones

Flint stones make a spark when they are struck. Heated stone stays hot for a long time. So some stones are used as cooking stoves.

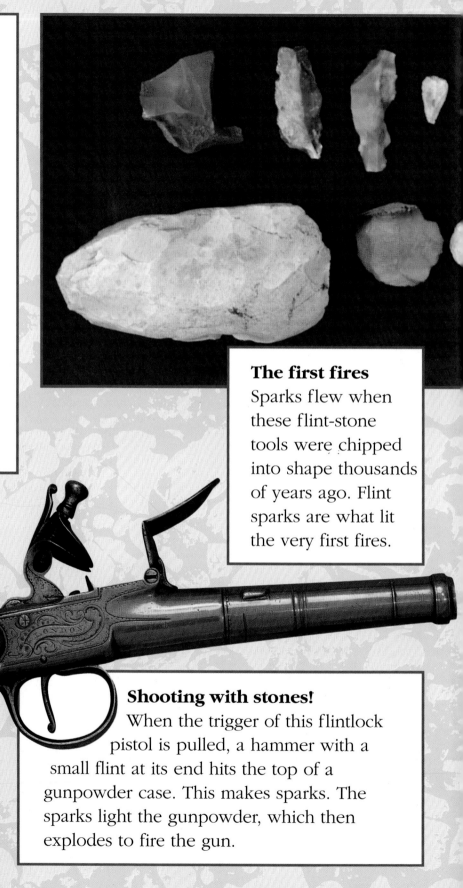

The first fires
Sparks flew when these flint-stone tools were chipped into shape thousands of years ago. Flint sparks are what lit the very first fires.

Shooting with stones!
When the trigger of this flintlock pistol is pulled, a hammer with a small flint at its end hits the top of a gunpowder case. This makes sparks. The sparks light the gunpowder, which then explodes to fire the gun.

Holding the heat

This stove in Senegal is made of three stones with a fire in the middle. The stones hold the heat and help the food cook evenly. Stones in deserts store heat from the sun. They can get so hot, that you can actually fry an egg on them!

Baking bread

Bread is being baked in hot stone ovens in Morocco. Bread ovens are often made of clay or fired bricks, too. In Turkey, hot stones are placed at the bottom of a deep oven that has been dug into the ground. The cooking pots are lowered into it. Similarly North American Indians used hot stones in a hollow log as an oven.

Sticky Stone

Sticky clay has been made over millions of years, as certain types of stone wear down into fine grains. Wet clay can be molded into all kinds of shapes. And when the shapes are baked dry, they stay that way!

China clay
Fine white china clay is dug out of the ground in Cornwall, England. It is used to make delicate china vases, tea sets, and ornaments.

Perfect pots
This Kenyan potter makes huge vases and water carriers. Later, they will be baked in a hot kiln. These tough earthenware pots need to be made from heavy gray or red clays. Gray clays contain lots of tiny bits of plant material. Red clays have iron in them.

Living in clay

In New Mexico, a fine house has been made from adobe. Adobe is a type of clay brick that has been dried in the sun. Clay is plastered over the brick walls as well.

Billions of bricks!

Most bricks are made of clay. These new bricks have been baked in a very hot kiln, in Bangladesh. They are likely to last longer than the sun-dried adobe.

Colorful clay

This Portuguese tile is a thin clay square. It has been baked twice. The second time was to give it its glassy coating, called glaze.

17

Paint It with Stone

Colored stones and clays have been used as paints for thousands of years.

Rock makeup

This Bangladeshi girl has used black kohl around her eyes to make them look more beautiful. Kohl comes from a type of rock crystal. It has been used as makeup for hundreds of years.

Red rock paintings

Long ago, Australian Aborigines painted the "lightning brothers" with paint made from red, earthy hematite rock. This is still used today in paints and crayons.

Beautiful blue

Four hundred years ago, a famous painter called Titian used lapis lazuli stone (left) to make a rich, blue paint. The color became known as Titian blue!

Yellow ocher walls

The yellow on these walls in Portugal comes from limonite, which is a kind of rust found in rocks and soils. Some rocks, like pyrolusite, are used as driers in paints. Ilmenite helps make the white smoke used when an airplane is skywriting!

Stones Do You Good

Some medicines contain chemicals that come from stones. Plants can be fed with certain stone chemicals to make them healthier and stronger.

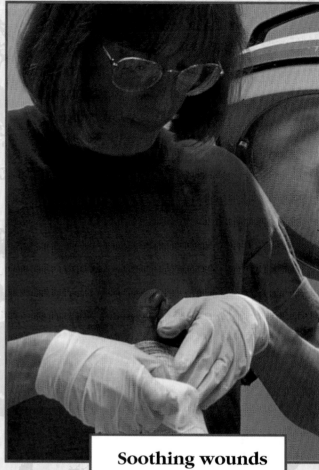

Fertile fields

In Bali, local farmers are digging up black lava sand to use as a fertilizer. Apatite crystals are also used. They contain phosphorus, which plants need in order to grow properly.

Soothing wounds

Sore skin and cuts can feel better with ointment on them. Some ointments contain sulfur taken from rock crystals. In the past, sulfur was used in medicines to treat infections as well.

A stream of sulfur

Bright yellow sulfur flows from a volcanic mountain in Sumatra. Sulfur is used in fertilizers as well as in medicines. It also helps get rid of fungi and molds that grow on plants.

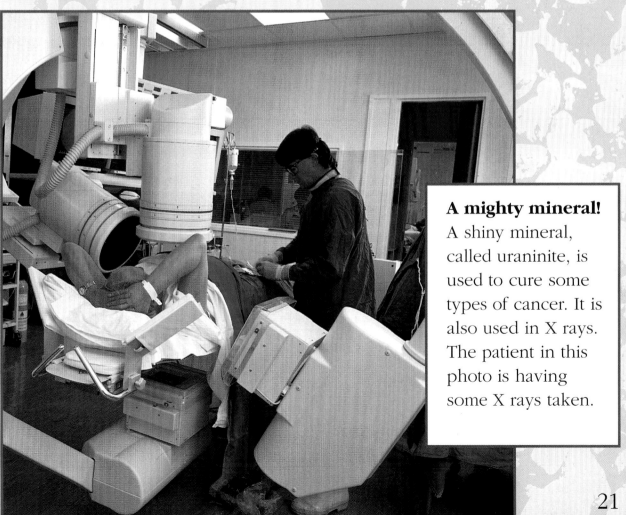

A mighty mineral!

A shiny mineral, called uraninite, is used to cure some types of cancer. It is also used in X rays. The patient in this photo is having some X rays taken.

Metal from Stones

Shiny, soft silver and hard iron can be found in stone. So can the slippery liquid, mercury!

Useful silver

Anklets and toe rings can be made from silver. So can cutting tools. Antiseptics often contain silver salts. Other silver chemicals help to make photographic film and paper.

Slippery mercury

Most mercury comes from red cinnabar rock. When the rock is melted, drops of silvery mercury form. Mercury is used in many thermometers, barometers, and, in small amounts, tooth fillings!

Rock solid!
San Francisco's famous Golden Gate Bridge is made of strong steel. Steel is mostly iron, a metal extracted from rocks. It has been used for tools and weapons for thousands of years.

Grains of gold
Gold is mined from rocks. But tiny amounts are worn away from rocks by water. People collect these grains in special sieves, called pans, like here in the Philippines.

Shimmering Stones

Dazzling diamonds and cool, blue sapphires can be discovered in stones. Gemstones are not only beautiful, they can also be very tough!

Gems in stone
These colored gems were found in gray stone, in Sri Lanka. The gems will be cut and polished to make them sparkle.

Shiny diamonds
You can only cut diamonds with other diamonds! This is because no other gem is as hard. Diamonds are also used as glass cutters, drill bits, and saws.

Gem jewelry

Gemstones are set into metal to make jewelry. This silver butterfly brooch is set with diamonds, rubies, and sapphires.

Ancient gems!

People say diamonds are forever—and all gemstones are very long-lasting! This French statue of Saint Foy is 1,000 years old. It is set with precious gems, like rubies, and semiprecious stones, such as amethysts.

Shifting Stones

What are deep, shifting sand dunes made from? Yes, stones! Sand is made over thousands of years from worn-down quartz, a type of stone. It is used in unbelievable ways!

Building with sand

Sand is mixed with cement and water to make mortar for building. The sticky mortar is used to hold bricks together.

Dazzling glass
This folding stained glass screen is used to cover a window. The glass in both the screen and the window is made mostly of melted grains of sand.

Glass shapes
A delicate oil jug is being made in Israel. Glass can be pressed and molded into shape. But it can also be shaped by blowing air into it through a long, iron pipe.

27

Find Out for Yourself

Find out more about stones for yourself with some arts and crafts activities. Here are some ideas to help you get started:

Searching for stones

Try making your own collection of stones. You will find lots of them where you live, or you could collect stones while on vacation. They will help you remember your visit. Look in flower beds, on a beach, a dried-up river, or at the edge of a plowed field. You can also find stones where there are bare rock faces—at the bases of cliffs, mountains, and on hillsides.

You may collect stones and pebbles just because they look pretty. A good way to display them is to cover them in water inside a glass bowl, so that they sparkle and shine.

If you want to know what types of stones you have found, there are plenty of books that can help you. A magnifying glass is useful, too. This will let you see all the colors in a stone and what it is made up of. For example, it could be grainy or contain crystals.

Painted stones

The shapes of some stones are like animals or objects. You can paint them with poster paints. In the picture, you can see a stone that looks like a ladybug. You could simply paint patterns on other stones that you find.

Lots of pots

It is fun to play with clay. You can buy it at craft stores. Make a simple thumb pot in your hands, like this blue-and-red one, or try a coil pot. Keep the clay damp with water while you shape it. Unless you fire them in a kiln, the pots won't be waterproof or very strong. But you can make them bright and colorful with poster paints.

A coil pot

Sprinkle a cutting board or plate with flour. Press some clay flat onto it to make the coil pot's base. Roll some more clay into a long, wormlike shape. Press the worm around the edge of the base. Make another worm, and coil it on top of the first one. Continue, until you are happy with the height and shape of the pot.

You can make different shaped pots by gently pushing the clay inward, or by starting with a different shaped base. If you want to, you can also paint your pot when the clay is dry.

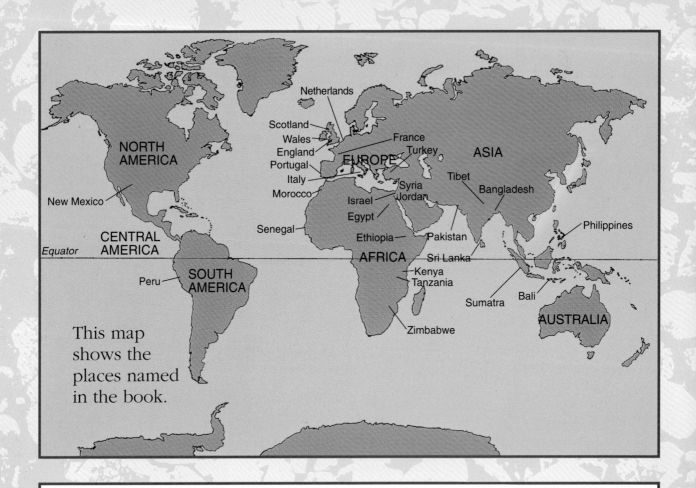

This map shows the places named in the book.

Further Reading

Benanti, Carol. *Gemstones*. Random Books, 1994.

Lye, Keith. *Rocks and Minerals*. Raintree Steck-Vaughn, 1993.

Russell, William. *Rocks and Minerals*. Rourke, 1994.

Index

Acknowledgments

Editors: Rachel Cooke, Kathy DeVico

Design: Neil Sayer, Joyce Spicer

Production: Jenny Mulvanny, Scott Melcer

Photography: Michael Stannard (pages 28 & 29)

For permission to reproduce the following copyright material, the author and publishers gratefully acknowledge the following:

Cover (top left) Marc Romanelli/The Image Bank, (top right) Bonhams London/The Bridgeman Art Library, (bottom left) Vaughn Fleming/Science Photo Library, (bottom right) Warwick Johnson/Oxford Scientific Films, (logo insert, front & back) J. H. Carmichael Jr./The Image Bank **title page** (main picture), Steve Alden/Bruce Coleman Limited (logo insert) J. H. Carmichael Jr./The Image Bank **page 6** (top) Carl Frank/Photo Researchers Inc./Oxford Scientific Films, (bottom) Andrew Davies/Bruce Coleman Limited **page 7** (top) Richard Packwood/Oxford Scientific Films, (bottom) Dan Gurarich/Photo Researchers Inc./Oxford Scientific Films **page 8** (left) Christine Osborne Pictures, (right) Musée Royale Afrique Centrale, Tervuren, Belgium/Werner Forman Archive **page 9** (top) Jeff Foott/Bruce Coleman Limited, (bottom left) Christine Osborne Pictures, (bottom right) Mike Birkhead/Oxford Scientific Films **page 10** (top) BSIP, Barrelle/Science Photo Library, (bottom) Sassoon/Robert Harding Picture Library **page 11** (top) Steve Turner/Oxford Scientific Films, (bottom) Christine Osborne Pictures **page 12** (top) Anna Barry/Eye Ubiquitous, (bottom) Ecoscene **page 13** (top) Nick Saunders/Panos Pictures, (bottom) Polypix/Eye Ubiquitous **page 14** (top) G. A. Maclean/Oxford Scientific Films, (bottom) The Victoria & Albert Museum/Robert Harding Picture Library **page 15** (top) Jeremy Hartley/Panos Pictures, (bottom) Lisl Dennis/The Image Bank **page 16** (top) Colin Molyneux/Bruce Coleman Limited, (bottom) Juliet Highet/The Hutchison Library **page 17** (top) Winkley/Ecoscene, (bottom left) Trygve Bølstad/Panos Pictures, (bottom right) Bill Hickey/The Image Bank **page 18** (top) David Constantine/Panos Pictures, (bottom) John Miles/Panos Pictures **page 19** (top) Francesco Turio Bohm/Galleria dell' Accademia, Venice/The Bridgeman Art Library, (insert, top) Vaughn Fleming/Science Photo Library, (bottom) F. Moura Mochado/The Image Bank **page 20** (top) John Greim/Science Photo Library, (bottom) Fritz Prenzel/Bruce Coleman Limited **page 21** (top) Andy Clarke/Science Photo Library, (bottom) Paul Van Reil/Robert Harding Picture Library **page 22** (top) Massimo Borchi/Bruce Coleman Limited, (bottom) J. C. Revy/Science Photo Library **page 23** (top) Steve Alden/Bruce Coleman Limited, (bottom) Chris Stowers/Panos Pictures **page 24** (top) Tim Page/Eye Ubiquitous, (bottom) Andrea Pistolesi/The Image Bank **page 25** (top) Bonhams, London/The Bridgeman Art Library, (bottom) Lauros-Giraudon/The Bridgeman Art Library **page 26** (top) Ray Ellis/Science Photo Library, (bottom) T. Souter/The Hutchison Library **page 27** (left) Christine Pemberton/The Hutchison Library, (right) Christine Osborne Pictures.